GALWAY COUNTY LIBRARIES
WITHDRAWN FROM CIRCULATION

Tree of Life

The Incredible Biodiversity of Life on Earth

Galway County Libraries

WRITTEN BY
Rochelle Strauss

ILLUSTRATED BY
Margot Thompson

A & C BLACK

For Oliver and Rosanne, with much love.

Acknowledgements

My deepest thanks to Valerie Hussey and Valerie Wyatt, whose faith and wisdom guided me in bringing this idea to life. Much gratitude as well to Margot Thompson for her breathtaking illustrations and Marie Bartholomew for crafting together concepts, words and images in such a beautiful way. Thanks also to Susan, Liz and Kate, for making me laugh when the numbers became overwhelming, and to Julia for helping me believe in my true calling as an "otter." Big hugs to my family, for their love and support. And a very special thank you to Rosanne, for constantly daring me to make my dreams come true.

Of course, this book would not have been possible without my intrepid team of technical reviewers: Joanne DiCosimo, Mark Graham and Robert Anderson, Canadian Museum of Nature; Ann Jarnet, Environment Canada; Liz Lundy, World Wildlife Fund; and Susan Gesner, Gesner & Associates Environmental Learning. Special thanks to them for their valuable input and insight.

J182,250
£19.50

A note on species and numbers

The sheer magnitude of biodiversity on Earth means that scientists do not know the exact numbers of species between and within the five kingdoms. The figures used in this book are based on their best estimates and have been rounded up or down to make them more manageable.

Published in 2005 by
A&C Black Publishers Ltd
37 Soho Square
London W1D 3QZ
www.acblack.com

Text © 2004 Rochelle Strauss
Illustrations © 2004 Margot Thompson

The rights of Rochelle Strauss and Margot Thompson to be identified as author and illustrator of this work respectively have been asserted by them in accordance with the Copyrights, Designs and Patents Act 1988.

ISBN 0 7136 7294 3

A CIP catalogue record for this book is available from the British Library.

Edited by Valerie Wyatt
Designed by Marie Bartholomew

Printed in China by WKT Company Limited.

A&C Black uses paper produced with elemental chlorine-free pulp, harvested from managed sustainable forests.

Published by permission of Kids Can Press Ltd, Toronto, Ontario, Canada. All rights reserved. No part of this publication may be reproduced in any form or by any means, graphic, electronic or mechanical including photocopying, recording, taping or information storage and retrieval systems – without the prior permission in writing of A&C Black Publishers Limited.

Kids Can Press acknowledges the financial support of the Government of Ontario, through the Ontario Media Development Corporation's Ontario Book Initiative; the Ontario Arts Council; the Canada Council for the Arts; and the Government of Canada, through the BPIDP, for their publishing activity.

Published in Canada by
Kids Can Press Ltd
29 Birch Avenue
Toronto, ON M4V 1E2

Published in the US by
Kids Can Press Ltd
2250 Military Road
Tonawanda, NY 14150

www.kidscanpress.com

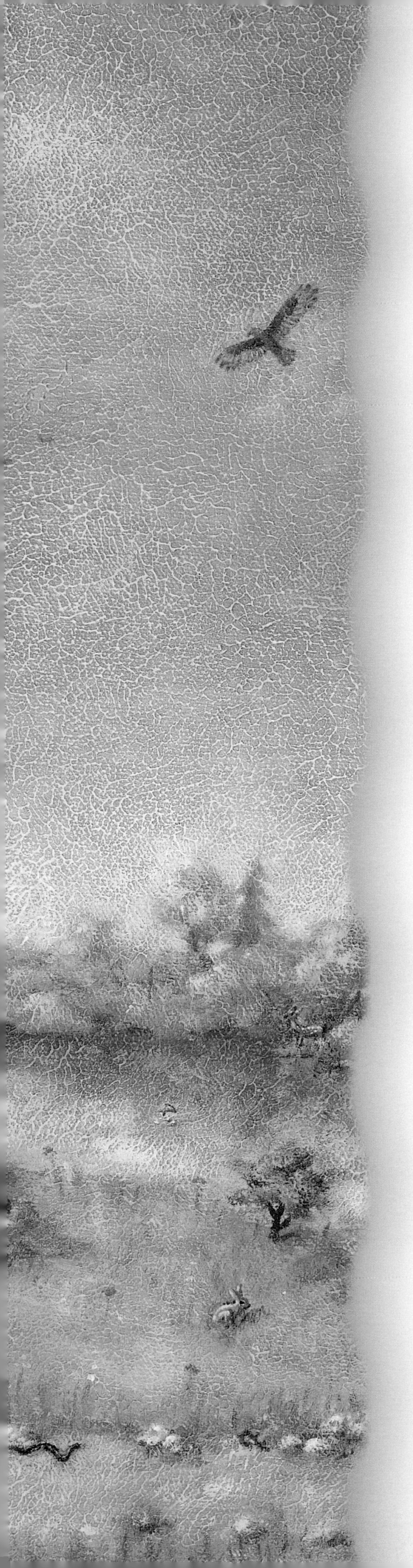

The Tree of Life

Do you have a family tree that shows how the members of your family – aunts, cousins, grandparents and so on – are related?

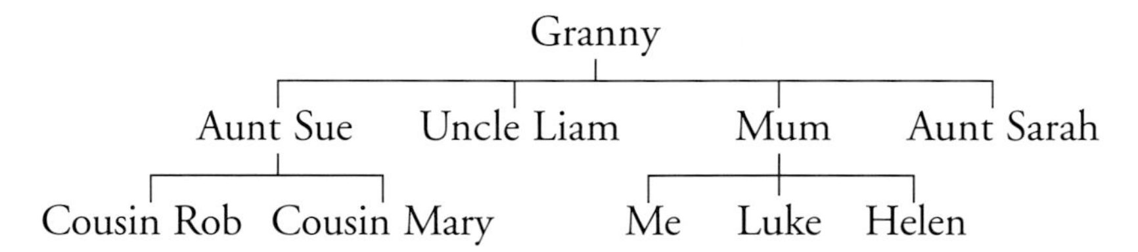

The Tree of Life is like a family tree for all living things. It shows us biodiversity, which is the incredible variety of life on Earth. It shows us how all living things are related – from bacteria too small to see with the naked eye to the largest mammal, the blue whale.

So far, scientists have discovered and named 1 750 000 different species. A species is a group of living things that share similar characteristics. Scientists do not agree on the exact number of species – there are just too many to be sure. The numbers in this book are based on their best estimates. If every species were represented by a leaf, there would be 1 750 000 leaves on the Tree of Life. All those leaves would be related, some closely and others more distantly.

Every part of the Tree of Life is important. A problem with one branch, one twig or even just one leaf may affect the whole tree. In this book we explore the Tree of Life and its branches. This will help us to understand our place within the Tree of Life and our impact on it.

The five branches

The Tree of Life is a way to organize, or classify, all living things. By organizing them into different branches, we can better understand how closely – or distantly – they are related.

The Tree of Life is often divided into five main branches called kingdoms.

Kingdom Monera
bacteria

Kingdom Fungi
mushrooms, toadstools, moulds, yeasts, mildews, etc

Kingdom Protoctista
paramecia, amoebas, algae, etc

Kingdom Plants
flowering plants, mosses, ferns, etc

Kingdom Animals
from invertebrates, such as sponges and spiders, to vertebrates, such as fish, amphibians, reptiles, birds and mammals

Each kingdom on the Tree of Life has a story to tell us about biodiversity and life on Earth.

Kingdom Protoctista
Kingdom Animals
Kingdom Plants
Kingdom Monera
Kingdom Fungi

Cyanobacteria are the oldest known bacteria. They are also the most important to life on Earth. More than 3.5 billion years ago, the first cyanobacteria began to create oxygen. This eventually allowed other life on Earth to exist.

Do you like beans? Or peas? Without bacteria, such as Rhizobium, *these legumes – and many other plants – couldn't survive. Some bacteria live in the roots of plants and help them to get the nutrients they need to survive.*

Monera

10 000 species

You can't see them, but they're out there – the 10 000 species of bacteria that make up the Kingdom Monera.

Bacteria are the smallest life forms on Earth. Each one is made up of just a single cell and is so tiny that 1000 of them would fit on the full stop at the end of this sentence. They're everywhere – on land, in water, even inside you. And there may be hundreds of thousands more species waiting to be discovered, even in the harshest environments on Earth – hot springs, sea vents and areas deep beneath the soil.

The Kingdom Monera accounts for less than 1 per cent of all species on the Tree of Life, but it is still an important kingdom. It contains the oldest living species. Fossils reveal that bacteria have been around for more than 3.5 billion years. These ancient bacteria were the basis of all life on Earth.

We sometimes think of bacteria only as carriers of disease, but bacteria are much more than that. Every living thing on the Tree of Life is descended from bacteria.

There are billions of bacteria in your intestines. One of them, Lactobacillus acidophilus, *helps to protect you from harmful bacteria.*

Monera – 10 000 leaves on the Tree of Life

Fungi

72 000 species

Would you recognize a fungus? You've probably seen or even eaten one today. Every time you bite into a piece of bread, you are eating yeast, a species of fungus. Scientists believe that as many as one million more fungi species have yet to be discovered.

Some fungi are parasitic – they grow on other living plants and animals and take their nutrients from them. But most fungi are decomposers – they take nutrients from dead plants or animals. Decomposers are the recyclers and cleaners on the Tree of Life.

Imagine a forest in the autumn, with billions of leaves falling to the ground. Where do all these leaves go? Fungi (and some bacteria) help to break them down and absorb them as food. And as they do this, they create carbon dioxide, which plants use to make their own food.

Without fungi, the Tree of Life would become buried under its own leaves.

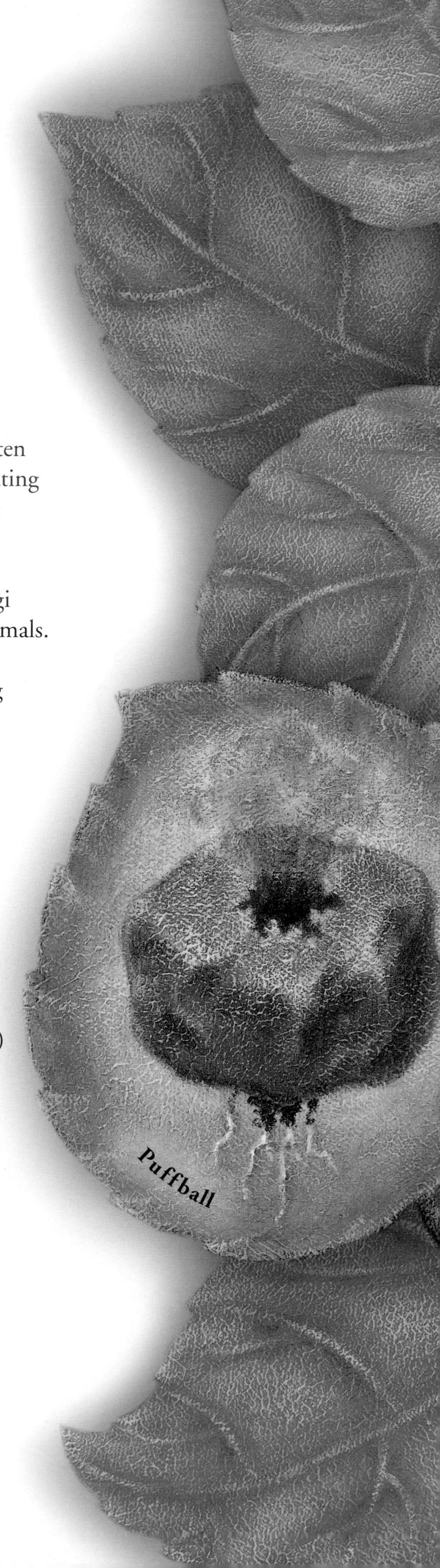

Fungi species

30 000 sac fungi (truffles, morels, yeast, lichen, etc)

22 250 club fungi (mushrooms, toadstools, puffballs, etc)

17 000 imperfect fungi (penicillin, candida, etc)

600 conjugation fungi (black bread moulds, etc)

... and others

Fungi – 72 000 leaves on the Tree of Life

Lichens are usually combinations of fungi and tiny plants. They are very sensitive to toxins (poisons) in their environment. When lichens start to die, it's an early warning signal that pollution levels may be high in that area.

One of the world's oldest and heaviest fungi is an Armillaria bulbosa *which lives beneath a forest floor in Michigan, in the United States. This 1500-year-old fungus may be heavier than an African elephant, but all you can see of it are tiny shoots poking up from the ground.*

The smallest puffball fungus is about the size of a chicken egg. The biggest is the size of a watermelon.

Most protoctista are microscopic, but some, such as seaweed, can be huge. The largest, the Pacific giant kelp, can grow up to 65 m long – as long as five buses parked end to end. Kelp provides shelter for marine animals.

Paramecia, a type of protozoa, have tiny hair-like structures called cilia. To help paramecia move in their watery habitats, the cilia move back and forth – a bit like rowing a boat.

The ocean's largest animal, the blue whale, feeds on its tiniest inhabitants, plankton.

Have you ever seen the surface of the ocean shimmer at night? You may be seeing fire algae, which produce light through bioluminescence, a chemical reaction.

Protoctista

80 000 species

Are they plants? Animals? The answer is yes ... and no. The Kingdom Protoctista (sometimes known as Protista) has a little bit of everything. Some species (the algae) are plant-like – they can make their own food. Others (the protozoa) are animal-like – they depend on other species for food.

Protoctista are found in water and other wet environments. In the ocean, they are a major food source known as plankton. Fish, shrimp and other crustaceans eat plankton and become food for other animals in turn, both in water and on land. This is called a food chain. When many food chains are linked together, they form a food web. Without plankton, many species would starve, and the food webs that support the Tree of Life might break down.

Algae have another important role. They help to maintain the balance of gases in the Earth's atmosphere. Algae, like plants, absorb carbon dioxide and use sunlight to create food for themselves. This process is called photosynthesis. By doing so, they also create the oxygen that all plants and animals need to survive.

Protoctista species

55 000 protozoa
(paramecia, amoebas, etc)

25 000 algae
(green algae, red algae, fire algae, etc)

Protoctista – 80 000 leaves on the Tree of Life

KINGDOM

Plants

270 000 species

Just about anywhere you look on land, you will find members of the plant kingdom. They range from the flowers in a window box to the trees that give us shade in the summer, and from the mosses you walk on in the woods to the vegetables you eat for dinner.

Plants provide valuable habitats for many animals. A habitat is an area where species can find the food, shelter, water and space they need to survive. Without habitats, animals could not survive.

Like protoctista, plants are also at the base of food chains. Plants make their own food – and become food for other living things. A rabbit nibbles a clover plant. A snake eats the rabbit, then a hawk eats the snake. When the hawk dies, bacteria and fungi feed on its body. Without plants, the food webs on the Tree of Life would collapse, and species would become extinct.

As plants make food for themselves during photosynthesis, they create oxygen. Next time you take a deep breath (or even just a tiny one), remember that the oxygen produced by plants makes life on Earth possible.

Plant species

235 000 flowering plants (maples, oaks, cacti, grasses, daisies, etc)

12 000 ferns (maidenhair fern, Boston fern, staghorn fern, etc)

10 000 mosses (peat moss, sphagnum moss, granite moss, etc)

630 conifers (pine trees, cedars, junipers, etc)

... and others

Plants – 70 000 leaves on the Tree of Life

High up in tropical rainforests, the bromeliad grows into a bowl of leaves attached to a tree. This bowl catches water and becomes a habitat for many species of frogs, insects, spiders and worms. The largest bromeliad is just a bit smaller than a rucksack. It can hold nearly 7.5 l of water.

The bee orchid is an excellent mimic. Its flower looks so much like a female bee that male bees are tricked into landing on it. When they fly away, they carry pollen with them to the next flower.

Milkweed is an important plant for monarch butterflies – it's where they lay their eggs and it's the main food for monarch caterpillars. Eating milkweed also makes the caterpillars and butterflies poisonous to other animals.

Many plants are used in medicines. The rosy periwinkle of Madagascar is a vital ingredient in two medicines that treat cancer. Rosy periwinkles are at risk because their habitat is disappearing.

Galway County Libraries

Stony corals are one of many coral species which build coral reefs. These reefs are the rainforests of the ocean – they provide a habitat for a huge diversity of animals.

A fruit-eating fish? The tambaqui of the Amazon River eats fruits that fall into the water. The seeds are dispersed in the waste of the fish.

Ants and acacia trees have a special relationship. Ants protect acacia trees from plant-eating insects, such as beetles and aphids. In return, the acacia tree provides nectar for the ants to eat and hollow thorns for shelter.

The Jamaican leaf-nosed bat helps to spread fig seeds. The bat eats the figs, then the seeds are dispersed in the bat's droppings.

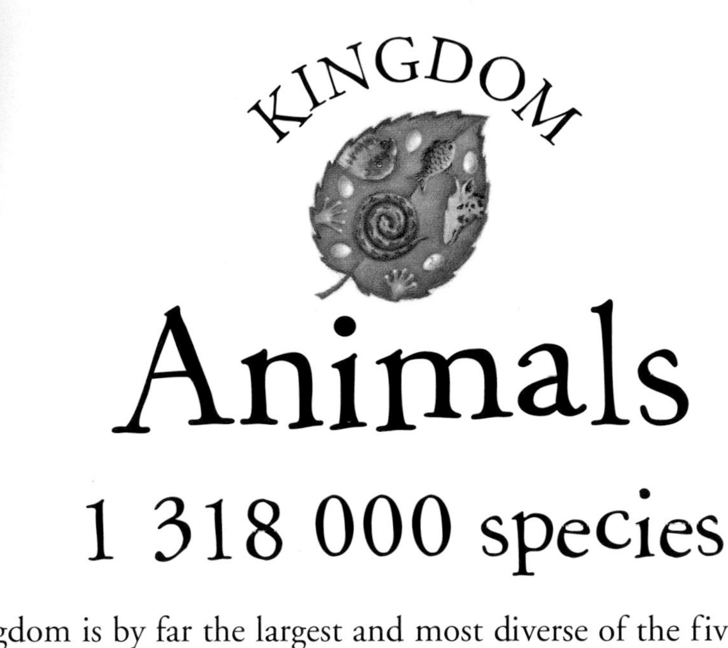

Animals

1 318 000 species

The animal kingdom is by far the largest and most diverse of the five kingdoms. It includes lions, tigers and bears, as well as jellyfish, sponges and sea slugs. In fact it accounts for about three-quarters of all leaves on the Tree of Life.

The species in this kingdom are further divided into two groups: invertebrates (animals without backbones, such as spiders, insects, sponges and worms) and vertebrates (animals with backbones – fish, birds, reptiles, amphibians and mammals).

With or without backbones, all animals share some characteristics. Unlike plants, animals cannot create their own food. They rely on other life forms for food. Some animals (herbivores) eat plants, while others (carnivores) eat the animals that eat the plants.

Plants rely on animals, too. Most flowering plants need animals, especially insects, to take pollen from one flower to another so that new seeds can form. Without animals, many plants could not produce seeds.

Animals also help to spread plant seeds. Birds and bats eat the fruit the plants produce. The fruit is digested and the seeds are dropped in the animal's waste, away from the adult plant. This gives the new plants room to grow. Squirrels and chipmunks collect seeds and bury them to eat later. The seeds that are forgotten grow into new plants.

Animals even help plants grow by providing much-needed nutrients. Animal dung is a rich fertilizer.

Animal species

1 265 500 invertebrates

52 500 vertebrates

Animals – 1 318 000 leaves on the Tree of Life

Animals → Invertebrates

1 265 500 species

Invertebrates live everywhere on Earth – on land and in water. They range from sea sponges, corals and jellyfish, to insects, spiders and worms. About the only thing they all have in common is that none of them has a backbone. Instead, many have an exoskeleton – a tough outer covering which protects them.

Of all the invertebrates, insects are probably the most familiar, because they make up more than three-quarters of all invertebrates.

Some invertebrates are enormous. The giant squid is probably the biggest invertebrate on the Tree of Life. It can grow up to 18 metres long and weigh more than 450 kilograms. Even its eyes are huge – as big as basketballs.

But most invertebrates are small enough to be easily overlooked. Their size makes them so difficult to find that scientists believe there may be millions more invertebrates still to be discovered and named.

Invertebrate species

950 000 mandibulates
(insects, centipedes, millipedes, etc)

75 000 arachnids
(spiders, ticks, mites, horseshoe crabs, etc)

70 000 molluscs
(snails, sea slugs, mussels, octopus, squid, etc)

40 000 crustaceans
(lobsters, crabs, crayfish, shrimp, barnacles, etc)

20 000 nematodes
(roundworms, etc)

16 000 annelids
(leeches, earthworms, etc)

9000 cnidarians
(jellyfish, corals, sea anemones, etc)

7000 echinoderms
(sea stars, sea urchins, sand dollars, etc)

5000 sponges

... and others

Invertebrates – 1 265 500 leaves on the Tree of Life

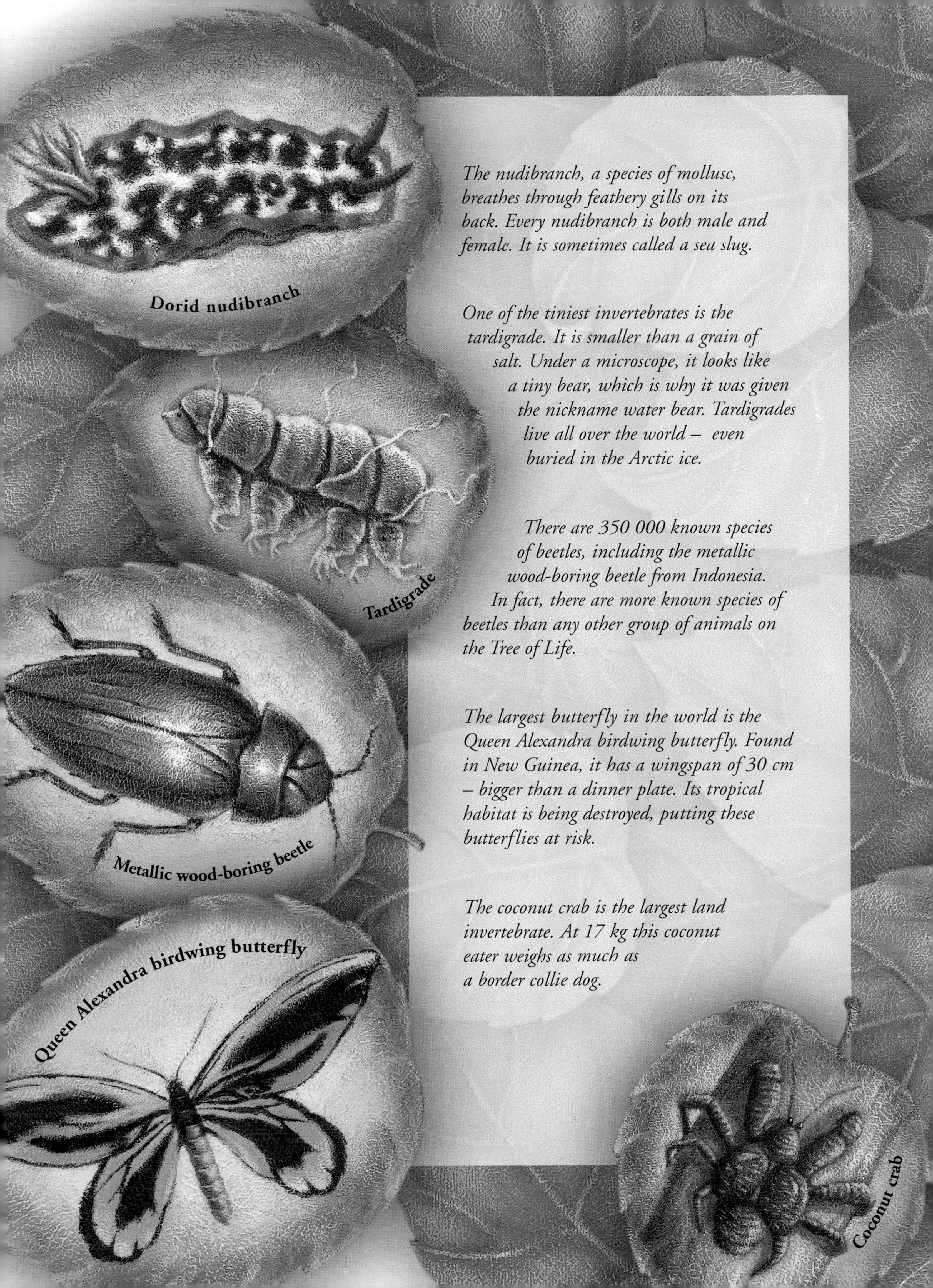

The nudibranch, a species of mollusc, breathes through feathery gills on its back. Every nudibranch is both male and female. It is sometimes called a sea slug.

One of the tiniest invertebrates is the tardigrade. It is smaller than a grain of salt. Under a microscope, it looks like a tiny bear, which is why it was given the nickname water bear. Tardigrades live all over the world – even buried in the Arctic ice.

There are 350 000 known species of beetles, including the metallic wood-boring beetle from Indonesia. In fact, there are more known species of beetles than any other group of animals on the Tree of Life.

The largest butterfly in the world is the Queen Alexandra birdwing butterfly. Found in New Guinea, it has a wingspan of 30 cm – bigger than a dinner plate. Its tropical habitat is being destroyed, putting these butterflies at risk.

The coconut crab is the largest land invertebrate. At 17 kg this coconut eater weighs as much as a border collie dog.

People thought the coelacanth had been extinct for 70 million years. Then in 1938, a live one was found. It was the first of many. Some scientists think this deep sea fish may be closely related to the first land vertebrates.

If you are ever strolling through a tropical rainforest, look up into the trees. You might spot an emerald tree boa, one of the few reptiles that bear live young instead of laying eggs.

Imagine a frog the size of a housefly! The strawberry poison dart frog may be small, but, like all dart frogs, it is poisonous. Its bright colours warn predators to stay away.

A toco toucan's bill may look heavy, but it is actually hollow – and it is surprisingly nimble. Ridges along the edge help the toucan hold and manipulate the fruit it eats.

Animals → Vertebrates

52 500 species

Vertebrates are the animals we know best. Why? Because they are easy to find, even without a microscope. You only need to look in a mirror to be face to face with one. Although vertebrates are all around us, they only make up a tiny portion of the species on the Tree of Life.

Vertebrates are divided into fish, birds, reptiles, amphibians and mammals. The main feature they have in common is a backbone, which is made up of a series of bones called vertebrae. The vertebrae hold and protect the spinal cord, which works with the brain to control everything in the body. Vertebrates also have an internal skeleton that allows movement and provides support and protection.

Today, vertebrates live on land and in water, but the first vertebrates lived only in the seas. About 360 million years ago, some of these early fish crawled out of the sea, and began vertebrate life on land.

J182, 250.

Vertebrate species

25 100 fish

9800 birds

8000 reptiles

4960 amphibians

4640 mammals

Vertebrates – 52 500 leaves on the Tree of Life

Yangtze River dolphin

Whales and dolphins, such as this Yangtze River dolphin, probably evolved from hoofed mammals. In other words, they started as land animals and moved into the sea. Today there are only 300 Yangtze River dolphins left in the wild.

Animals→Vertebrates→Fish

25 100 species

Fish are aquatic – they spend their lives in water. Most fish live in either freshwater or saltwater. But some, such as eels and salmon, spend part of their life in both.

Fish are the most diverse vertebrate species. They are made up of three groups – bony fish, cartilaginous fish and jawless fish.

Bony fish are the most common. As their name suggests, they have a full skeleton made of bone. Trout, salmon, catfish and goldfish are examples of bony fish.

Cartilaginous fish have a skeleton made of cartilage – like the stuff your nose is made of. They include sharks and rays.

Jawless fish, such as lampreys and hagfish, are the rarest and most primitive. Lampreys have a sucker-like mouth, which they attach to other fish. They eat by scraping away the flesh of these fish with a ring of sharp teeth. Hagfish are scavengers which feed on dead and dying fish.

Most fish are cold-blooded – their body temperature is the same as their surroundings. Most have gills, as well as scales and fins which help them to swim.

Fish are an important food source on the Tree of Life. Many top predators, such as grizzly bears, eagles, sharks and even people, are fish-eaters.

Fish species

24 150 bony fish
875 cartilaginous fish
75 jawless fish

Fish – 25 100 leaves on the Tree of Life

African lungfish

The whale shark is not only the biggest shark, it is also the biggest fish in the sea. At up to 18 m long, it is more than 100 times longer than one of the smallest sharks – the dwarf dogshark. Surprisingly, it eats some of the smallest life on the Tree of Life – plankton.

The clownfish lives among the poisonous tentacles of the sea anemone, but has nothing to fear. The anemone protects the clownfish from predators and, in return, gets food scraps that the fish leaves behind.

The American eel begins its life in the Sargasso Sea, near Bermuda. As it grows older, it migrates long distances to freshwater lakes, streams or coastal areas. It returns to the Sargasso Sea at the end of its life to breed.

When lakes and rivers dry up, the African lungfish buries itself in the mud and waits for rain. While buried, it uses its mouth and lungs, instead of its gills, to take in air.

Galway County Libraries

Animals→Vertebrates→Birds

9800 species

Believe it or not, birds are descendants of dinosaurs. How do we know? By comparing their skeletons. Birds have the same type of ankle joint and hips as dinosaurs. Birds are also closely related to reptiles. They have scales on their legs and their feathers probably evolved from scales, too. But unlike reptiles, birds are warm-blooded – they maintain a constant body temperature.

Besides feathers, birds' most obvious features are wings and beaks. Wing shapes tell us a lot about how birds live. Short wings are good for twisting through forests. Long, narrow wings are great for soaring on air currents, and flipper-like wings are ideal for diving and swimming.

A bird's beak gives us clues about what it eats. Birds of prey have strong, hooked beaks to tear into their prey. Other birds have short, thick, curved beaks to crack open seeds and nuts. Nectar sippers have long, straw-like beaks.

Birds are the great migrators on the Tree of Life. Almost half of all birds migrate. Some fly great distances to and from breeding areas to escape cold climates or to find food. Migrating birds depend on many habitats along the way as resting places. If even one habitat on their route changes, this can spell disaster for these migrators.

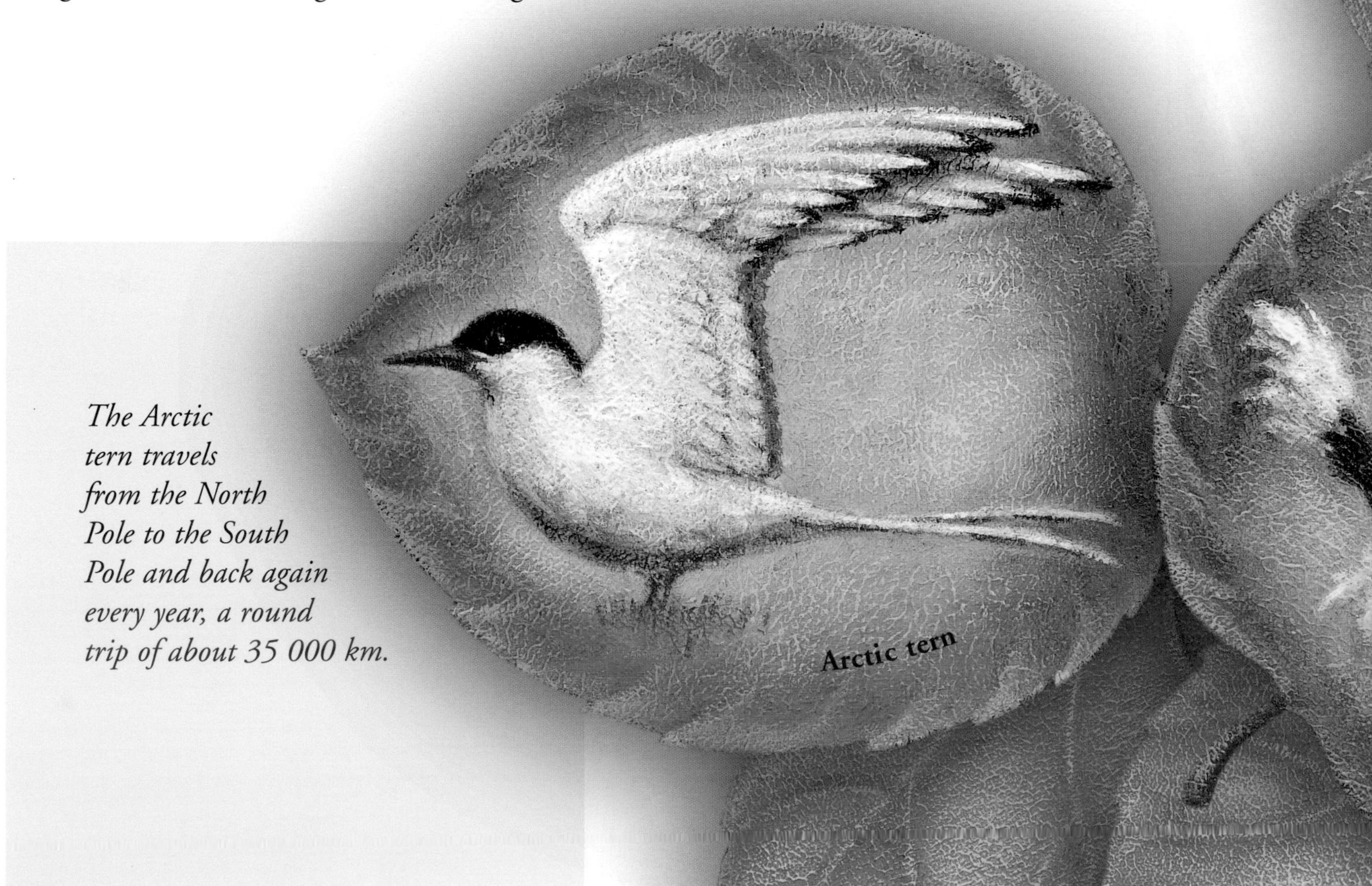

The Arctic tern travels from the North Pole to the South Pole and back again every year, a round trip of about 35 000 km.

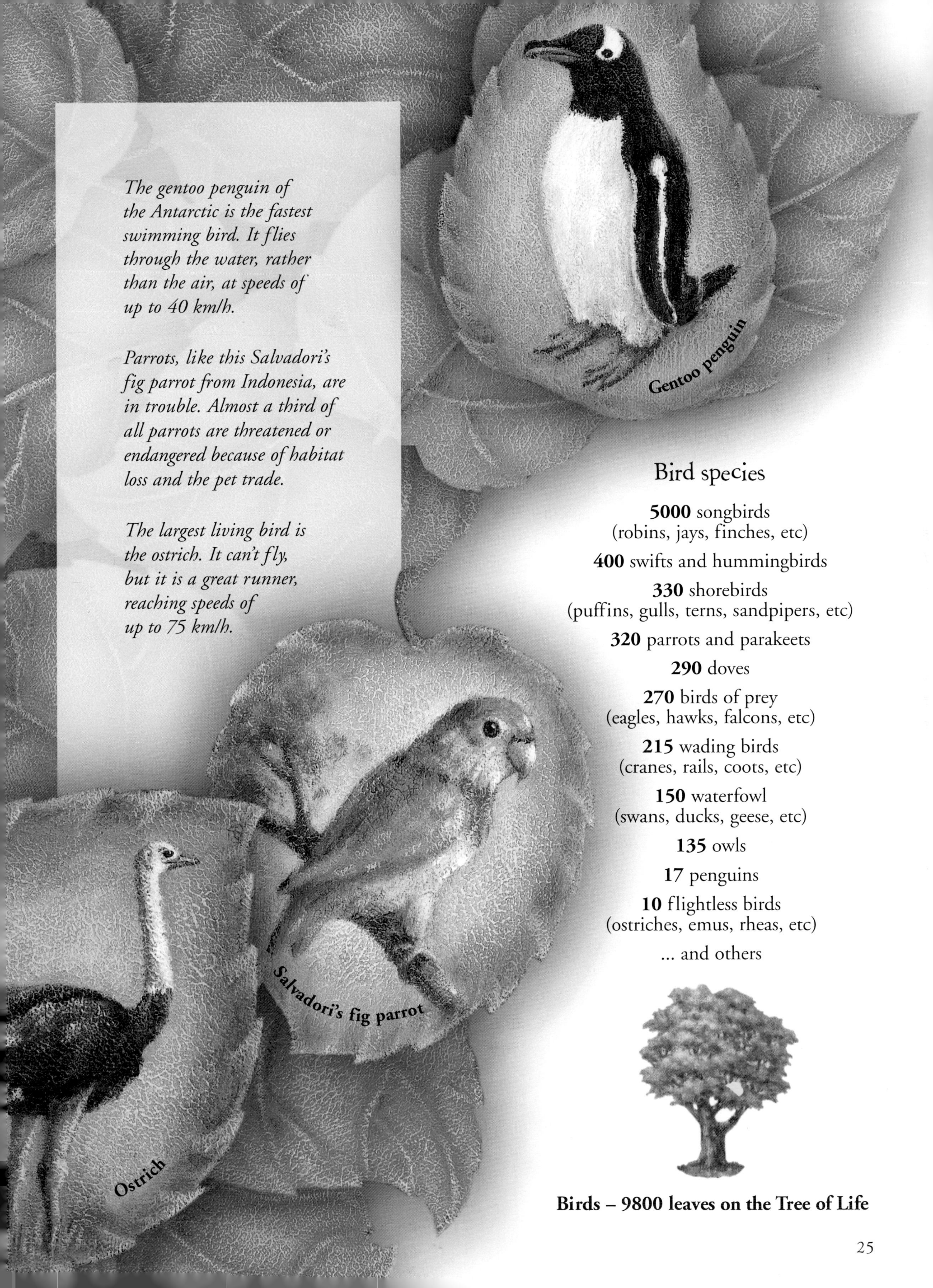

The gentoo penguin of the Antarctic is the fastest swimming bird. It flies through the water, rather than the air, at speeds of up to 40 km/h.

Parrots, like this Salvadori's fig parrot from Indonesia, are in trouble. Almost a third of all parrots are threatened or endangered because of habitat loss and the pet trade.

The largest living bird is the ostrich. It can't fly, but it is a great runner, reaching speeds of up to 75 km/h.

Bird species

5000 songbirds
(robins, jays, finches, etc)

400 swifts and hummingbirds

330 shorebirds
(puffins, gulls, terns, sandpipers, etc)

320 parrots and parakeets

290 doves

270 birds of prey
(eagles, hawks, falcons, etc)

215 wading birds
(cranes, rails, coots, etc)

150 waterfowl
(swans, ducks, geese, etc)

135 owls

17 penguins

10 flightless birds
(ostriches, emus, rheas, etc)

... and others

Birds – 9800 leaves on the Tree of Life

Animals→Vertebrates→Reptiles

8000 species

Slimy? No! Reptiles have dry, scaly skin, like their famous prehistoric cousins, the dinosaurs. Their scales help to trap moisture in their bodies so they don't dry out. That's why reptiles are so successful in desert habitats. But reptiles are not just desert dwellers – they can be found in a range of habitats, from land to freshwater and even in the oceans.

Scales are not the only thing reptiles have in common. They are also cold-blooded (that's why reptiles bask in the sun – to warm up), and most lay eggs, rather than bear live young.

Reptiles are divided into five groups – lizards, snakes, turtles and tortoises, crocodiles, alligators and caimans, and tuatara. All play an important role in the Tree of Life as both predators and prey. Most reptiles are carnivores (animal-eaters), although some lizards are herbivores (plant-eaters). There are also omnivores, such as turtles. They eat both animals and plants.

Reptile species

4320 lizards

3300 snakes

350 turtles and tortoises

28 crocodiles, alligators and caimans

2 tuatara

Reptiles – 8000 leaves on the Tree of Life

Panther chameleon

Tuataras live only in New Zealand. They are the most ancient type of reptile alive today. They are directly related to an early kind of reptile which roamed Earth before the dinosaurs. Tuataras have changed little since that time, but today they are very rare.

The green sea turtle spends most of its life in the ocean. Every two or three years, it migrates almost 2000 km back to the beach where it hatched, to mate and lay eggs. Like all sea turtles, green sea turtles are at risk.

An African rock python can grow up to 8.5 m long – which makes it big enough to eat an antelope! These pythons also eat pigs, baboons and monkeys.

Madagascar is home to almost half the world's chameleons, including the panther chameleon. This insectivore (insect-eater) can flick out its sticky tongue to almost twice the length of its body to grab an unsuspecting insect.

During hibernation or when food is scarce, the gila monster can live off the fat in its tail. Gila monsters are at risk and may soon become endangered due to habitat loss and the pet trade.

The Brazilian horned frog lives in the rainforests of Brazil and Argentina. It is a voracious eater and will devour just about anything it can catch. It swallows its food whole – including small birds, rodents and other frogs.

The female Surinam toad of South America carries her young on her back. The eggs are deposited in tiny holes on her back. Her skin swells up around the eggs to protect them. They stay there until the young toads emerge and swim away.

Most salamanders are about as long as a pen, but the Japanese giant salamander grows to ten times that length. These giants spend their lives in water and are nocturnal (active at night). They eat crabs, fish and other small amphibians.

The red-spotted newt of North America begins life in the water, metamorphoses and moves on to land. When it is old enough to breed, it moves back into the water.

Animals→Vertebrates→Amphibians

4960 species

Amphibians are the only vertebrates that go through metamorphosis – the complete change from one form to another. Frogs, for example, lay their eggs in water. Tadpoles emerge from the eggs and live in the water, using their gills to breathe. As they grow, they begin to change. Lungs and legs develop. Tails shrink. When metamorphosis is complete, their lungs and legs make them ready for life on land.

There are three groups of amphibians – frogs and toads, salamanders and newts (amphibians with tails), and caecilians (legless, wormlike amphibians). Frogs and toads are by far the largest group of amphibians.

Amphibians have an important part to play on the Tree of Life. As tadpoles, they are eaten by fish, birds, reptiles, mammals and even insects, such as the larvae of dragonflies. As adults, they are still prey for some animals, but they also become predators. They eat insects, worms and fish. Some even eat rats and ducklings.

Amphibians are also important as indicator species (species whose health indicates the state of the environment). Their thin, slimy skin absorbs water and air, making them sensitive to both water and air pollution. They are also vulnerable to ultraviolet (UV) light. A drop in amphibian numbers is one of the first signs that something is wrong in the environment.

Red-spotted newt

Amphibian species

4400 frogs and toads

400 salamanders and newts

160 caecilians

Amphibians – 4960 leaves on the Tree of Life

Animals → Vertebrates → Mammals

4640 species

Mammals are one of the smallest groups on the Tree of Life, yet they are found in almost every environment – land, water and air. Yes, air. Bats are mammals, and the only ones capable of true flight.

In size and shape, mammals are quite diverse. One of the smallest, the bumblebee bat from Thailand, weighs only about as much as two jellybeans. The largest mammal, the blue whale, weighs in at about 180 tonnes.

Mammals can be divided into three main groups. Most mammals are placental and give birth to live young. Some, such as kangaroos, are marsupials and carry their young in a pouch. The monotremes, such as the echidna, or spiny anteater, lay eggs.

What do such different mammals have in common? One thing is hair. All mammals have hair at some point in their lives, whether they have just a few whiskers or fur which covers their whole body. Even dolphins have a few bristles near their snout. Hair provides warmth for most mammals. It also helps to camouflage them. Spots, stripes and colours allow mammals to blend in with their environment or each other. Mammals also have mammary glands and feed milk to their young.

Mammals are connected by food chains and webs, like other species on the Tree of Life. Often the number of one mammal species is affected by the number of another species, so the two are in balance with each other.

The wombat is one of the largest burrowing mammals, and is slightly larger than a bulldog. This marsupial's pouch faces backwards so its young are protected from dirt as it burrows.

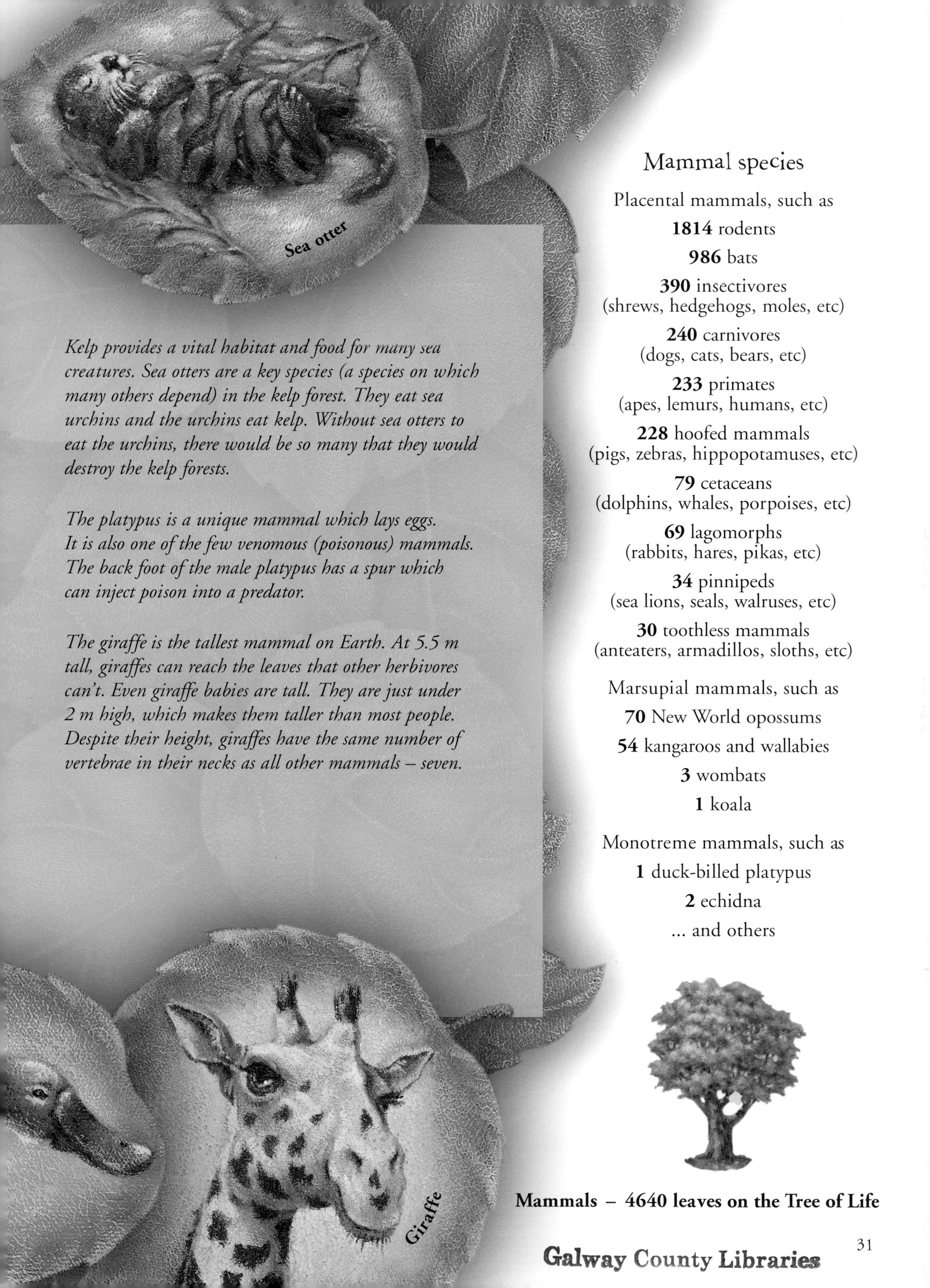

Kelp provides a vital habitat and food for many sea creatures. Sea otters are a key species (a species on which many others depend) in the kelp forest. They eat sea urchins and the urchins eat kelp. Without sea otters to eat the urchins, there would be so many that they would destroy the kelp forests.

The platypus is a unique mammal which lays eggs. It is also one of the few venomous (poisonous) mammals. The back foot of the male platypus has a spur which can inject poison into a predator.

The giraffe is the tallest mammal on Earth. At 5.5 m tall, giraffes can reach the leaves that other herbivores can't. Even giraffe babies are tall. They are just under 2 m high, which makes them taller than most people. Despite their height, giraffes have the same number of vertebrae in their necks as all other mammals – seven.

Mammal species

Placental mammals, such as

1814 rodents

986 bats

390 insectivores
(shrews, hedgehogs, moles, etc)

240 carnivores
(dogs, cats, bears, etc)

233 primates
(apes, lemurs, humans, etc)

228 hoofed mammals
(pigs, zebras, hippopotamuses, etc)

79 cetaceans
(dolphins, whales, porpoises, etc)

69 lagomorphs
(rabbits, hares, pikas, etc)

34 pinnipeds
(sea lions, seals, walruses, etc)

30 toothless mammals
(anteaters, armadillos, sloths, etc)

Marsupial mammals, such as

70 New World opossums

54 kangaroos and wallabies

3 wombats

1 koala

Monotreme mammals, such as

1 duck-billed platypus

2 echidna

... and others

Mammals – 4640 leaves on the Tree of Life

Galway County Libraries

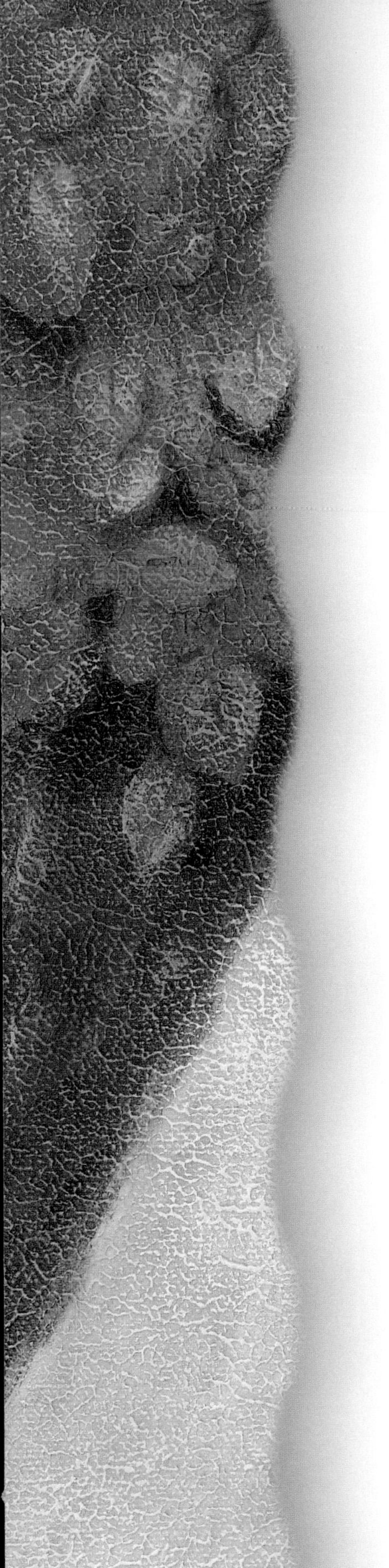

Animals → Vertebrates → Mammals → Primates → Humans

One species

Humans are one of the 233 species of primates.
We are closely related to the great apes (chimpanzees, gorillas and orangutans).

We are one of the 4640 species of mammals.
We have hair, mammary glands and give birth to live young,
as most mammals do.

We are one of the 52 500 species of vertebrates.
We have a backbone to protect our spinal cord, like all other vertebrates.

We are one of the 1 318 000 species of animals.
We breathe air and rely directly or indirectly on plants for food,
like all other animals.

Humans are one of the 1 750 000 species on the Tree of Life.
We are one leaf on the Tree.

Yet there are more than six billion humans on Earth, which means
that we have the greatest impact on the Tree of Life.

Humans – 1 leaf on the Tree of Life

Changes to the Tree of Life

The Tree of Life is constantly changing. New species are discovered every day, and known species are sometimes reclassified. Scientists think there may be as many as 20 million species on the Tree of Life.

Species are also being lost at an alarming rate. As many as 27 000 species may be lost every year – that's 74 species per day, 3 species per hour.

Every species, from fire algae to mangrove trees to zebras, needs a habitat that provides food, water and shelter. Cutting down forests, draining wetlands and ploughing grasslands to build motorways, roads, farms, towns and cities reduces habitat. Air, soil and water pollution threatens the quality of the habitats that remain.

- An area of tropical rainforest equivalent to two football fields is cut or burned down every second to make room for fields and farms. Without the habitat tropical forests provide, more than half the world's plant and animal species may be lost.

At risk

25 971 plants
1192 birds
1137 mammals
938 molluscs
752 fish
555 insects
408 species of crustaceans
296 species of reptiles
157 species of amphibians

- Almost half of all the mangrove forests have been cut down for timber, or damaged by pollution, fishing and urban growth. Without mangrove forests, coastlines may slip into the ocean. Animals will lose vital habitat.

- Almost a third of coral reefs have been lost, due to development, overfishing and pollution. Scientists estimate that half will be gone by 2010 at the current rate.

- Half the wetlands in the United States and many in Canada have been drained to make room for farms, houses and factories.

The extinction of even one species weakens the chain of connections between all the species on the Tree of Life. If too many sea otters die, sea urchins will destroy the kelp forests. If milkweed plants die, monarch caterpillars will starve. If plankton dies, whole marine food webs will collapse.

The maths is simple. Loss of habitat equals loss of species. And the loss of even one species on the Tree of Life affects all species.

Rainbow parrotfish

Some species at risk

Red panda
Maidenhair tree (gingko)
Giant clam
Fiji banded iguana
Wood poppy
Golden toad
Seaside centipede lichen
Macaroni penguin
Rainbow parrotfish
Sumatran orangutan
Lined seahorse
Chinese egret

Becoming guardians of the Tree of Life

The Tree of Life is not here for us to prune – to cut off a branch, to trim a twig. It is not something we can dig up and plant somewhere else. We must learn to live our lives within the Tree – as one leaf among 1 750 000. We are part of the Tree of Life. We are its guardians, not its gardeners.

When we forget that we are part of the Tree of Life, our impact can be damaging, but when we remember, we can be incredibly powerful. It doesn't take much to become a guardian of the Tree of Life. You don't need a lot of money or even a lot of time. All you need is the desire, determination and courage to think a bit differently. Here are just a few simple things you can try on your own, with your family or as a class.

Learn more Through understanding biodiversity and the connections between everything better, you can make choices in your life that help protect the Tree of Life. Start by choosing just one species from this book. Learn everything you can about it. Then learn about the species that interact with your species. Now look at the things you do in your everyday life. How do your actions have an impact on these species?

Reduce your impact on the Tree of Life Think about how you help or hinder biodiversity. Do you drive to school rather than walk? Cars add pollution to the air. Do you recycle waste, or turn off the lights when you leave a room? If so, you are saving resources and helping the environment, which in turn helps biodiversity. Count up all the things you do in a month that protect the Tree of Life. Try to add at least one new thing to that list every month.

Create a wildlife habitat In a house or a flat or even at school, you can help to create habitats. Fill window boxes with plants which attract insects and birds. Let your lawn grow and add some native plant species to attract wildlife to your garden. Plant a butterfly garden in your school grounds.

Have a clean-up Involve your school or community in cleaning up a local park or an area of wasteland. City habitats are vital to wildlife. Invite local media and encourage community businesses to become involved.

Educate others Start a biodiversity club at school or in your community. Write a newsletter or hold an awareness day to help others learn more about biodiversity.

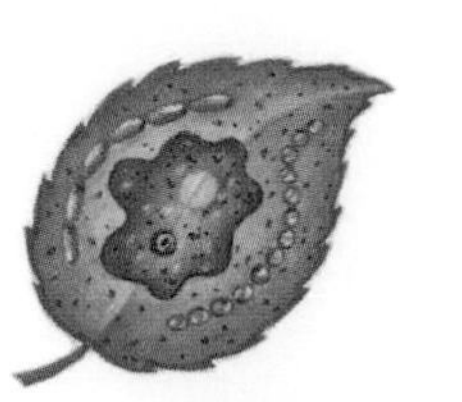

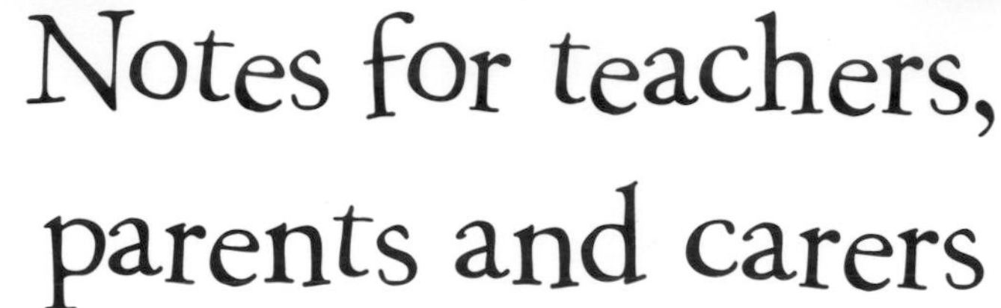

Notes for teachers, parents and carers

Classification and the Tree of Life
Classification is a system of sorting and grouping living species using the characteristics they have in common, such as their shape, how they move and how they reproduce and develop. Once classified, they can be given a place on the Tree of Life based on their connection with other species.

Taxonomy is the science of studying, identifying and classifying living organisms. It has fascinated humans for thousands of years. Early on, living things were sorted into two categories – plants and animals. When scientists discovered microscopes, they observed bacteria and protoctista for the first time. This led to the realization that two kingdoms were not enough. The Kingdom Monera was created, containing both bacteria and protoctista.

In the mid-1950s, protoctista (celled organisms with a nucleus) were given their own kingdom. Monera remained a kingdom containing only bacteria (single-celled organisms without a nucleus). A four-kingdom system was born – plants, animals, monera and protoctista.

The current five-kingdom system was established a short time later. Fungi, though grouped with plants, are really quite unlike plants. (While plants can make their own food, fungi cannot.) Yet fungi are not animals, either. And they certainly are not single-celled organisms. And so fungi became the fifth kingdom.

Recently, new technologies have allowed us to take a closer look at the genetic make-up of plants and animals. New discoveries are causing us to rethink former groupings. Today, scientists are looking at the possibility of dividing Monera into two kingdoms. They are also debating the role of viruses and whether they should have their own kingdom. Before long, the Tree of Life may grow to six or even seven kingdoms.

Biodiversity and the Tree of Life
Simply put, biodiversity is the incredible variety of life on Earth – from the variety of species, to the variety within species, to the variety of ecosystems (communities of living things and habitats linked together).

Most of us are aware of this diversity. We see it in everyday life. We know about different habitats – forests, oceans, grasslands, deserts. We recognize different species – black bears, polar bears, sun bears. And we may even be aware of differences within species – black bears can be black, brown or even cinnamon in colour.

Biodiversity is not just about individual species or individual ecosystems. It is also about how everything on the Tree of Life interacts. Everything is interconnected in some way. Plants rely on animals for pollination, animals rely on plants for oxygen and food, and fungi and bacteria break down their waste.

Pressure on one species – or one ecosystem – can have a big effect on all living things. If relationships are jeopardized as a result of the loss of species or altered habitat, the damage may be dramatic and irreversible.

From a human perspective, biodiversity is vital to our existence. The potential for new foods and medicines from the natural world is huge. Only a fraction of all the plant species on the Tree of Life are currently used for food, while as many as a quarter have already proved to have medicinal value. Yet our activities are destroying species and spaces faster than we can assess their potential.

Biodiversity is important for another reason. Every leaf on the Tree of Life, every species and every individual, has its own intrinsic value, independent of any value we place on it. Most children recognize this. Many have their own inherent connection to nature. They are born with a sense of knowing that other species are alive and should be treasured. If you have ever watched young children interact with an animal or experience nature, either in a natural setting or in a museum, or even while watching a video, you have seen their eyes light up with wonder. This appreciation is often lost as they grow up.

Our role as parents and teachers is to help nurture and foster this wonder and to channel it towards a greater sense of awareness, responsibility and stewardship. We need to foster a biodiversity ethic that recognizes the diversity and interconnection of all species and the role every species plays within the Tree of Life. This biodiversity ethic will help children understand their place within the Tree, as one of its many leaves. It will help children to see themselves as guardians and guide them to take action now and as adults to conserve biodiversity. That's what this book is about. It is one step in helping to foster a biodiversity ethic – but the next steps are yours.

What can you do?

- Promote the wonder of nature at home and in the classroom. Surround yourself, your family or your pupils with books, magazines and videos about the natural world. Go on field trips to parks, conservation centres, local woodlands and museums. Volunteer with an environmental group for a clean-up or fundraising event.

- Incorporate biodiversity and the Tree of Life into everyday events. At home, invite your children into the garden with you. As you plant, talk about the life cycles of plants, seasonal changes and the relationship between plants and wildlife. Engage children in discussions about purchases, and explore more environmentally responsible choices (less packaging, recyclable containers, alternatives to chemical cleaners and pesticides and so on). In the classroom, work nature into everything you study. Maths problems can revolve around numbers of species, writing projects can explore biodiversity issues, and art projects can involve nature in many ways.

- Start a dialogue about nature and the environment which allows children to explore and experiment with their own thoughts and feelings. Ask them what biodiversity means to them. Choose a topical biodiversity issue and have a debate. This will help children develop their own voices and opinions on issues.

In a world that is more and more removed from nature, we have to make an extra effort to ensure our children remain connected to the natural world. Making these connections at an early age helps a child develop a lifelong passion for nature and the Tree of Life.

Index